Aircraft Spin Tutorial

Author

Wilbur Hankey—AIAA fellow

Coauthor

Stephen Scurria

AuthorHouse™ LLC
1663 Liberty Drive
Bloomington, IN 47403
www.authorhouse.com
Phone: 1-800-839-8640

© 2013 Dr. Wilbur Hankey & Stephen Scurria. All rights reserved.

No part of this book may be reproduced, stored in a retrieval system,
or transmitted by any means without the written permission of the author.

Published by AuthorHouse 10/04/2013

ISBN: 978-1-4918-1679-0 (sc)
ISBN: 978-1-4918-1680-6 (e)

Any people depicted in stock imagery provided by Thinkstock are models,
and such images are being used for illustrative purposes only.
Certain stock imagery © Thinkstock.

This book is printed on acid-free paper.

Because of the dynamic nature of the Internet, any web addresses or links contained in this book may have changed since publication and may no longer be valid. The views expressed in this work are solely those of the author and do not necessarily reflect the views of the publisher, and the publisher hereby disclaims any responsibility for them.

Aircraft Spin Tutorial

Author

Wilbur Hankey - AIAA fellow

Coauthor

Stephen Scurria

Nomenclature

A - aerodynamic force vector

b - wing span

c - mean aerodynamic chord

C_l, C_m, C_n - moment coefficients about roll, pitch and yaw axes

D' - aerodynamic drag in cylindrical axis

e_1, e_2, e_3 - unit vectors in vertical cylindrical axes

g - acceleration of gravity

H - angular momentum vector

I_x, I_y, I_z - principal moments of inertia about body axes

i, j, k - unit vectors for body axes

i_o - dimensionless moment of inertia

L' - lift force in cylindrical axis

L, M, N - aerodynamic moments about body axes in roll, pitch and yaw

p, q, r - angular velocity about body axes

q* - dynamic pressure

R - spin radius

R_o - dimensionless radius

V - flight velocity

V_o - dimensionless velocity

Y' - side force in cylindrical axis

$\dot{z}$ - sink rate

α - angle of attack

β - side slip angle

ρ - density

ϕ - roll angle

θ - pitch angle

ω - angular spin velocity

ω_o - dimensionless spin velocity

Introduction

Aircraft spin has received considerable attention throughout the history of aviation. In the early days of flight, little was known about the cause. The Wright brothers for the first time conceived the idea of roll control to stabilize the aircraft and prevent the entry into spin. Flight experiments helped provide scientific information and a procedure for spin recovery was eventually developed (Ref1). Vertical wind tunnel experiments with dynamically similar models produced useful spin data. Also wind tunnel experiments with rotary balances were used to obtain the required aerodynamic coefficients on a stalled aircraft in spin (Ref2). Today CFD is capable of evaluating the aerodynamic coefficients for a stalled spinning aircraft (Ref3). But aircraft spin has continued to plague pilots. It is still a perplexing and dangerous situation encountered in flight.

More F-4 aircraft were lost during the Vietnam conflict due to spin than in combat. The two-seater aircraft usually had one fatality for each aircraft lost. It is important for all aeronautical engineers to possess an understanding of aircraft spin. Aircraft spin is a stable autorotation condition in which the plane descends in a dangerous helical path. The vehicle is stalled, the power is out and the control surfaces are ineffective. Because the condition is a stable one, the pilot has difficulty escaping. The stable equilibrium condition is similar in concept to the stable equilibrium cruise state. The major difference is that the aerodynamic moments are zero during cruise while the aerodynamic moments in spin are balanced by the gyroscopic dynamic moments to produce an equilibrium state.

Aircraft spin occurs in a power off, stalled condition. The vehicle rolls and rotates in a nose down position. Eventually the aircraft reaches a steady state position with constant roll angle (ϕ), pitch down angle (θ) and a constant spin rate (ω). The sink rate ($\dot{z}$) is constant and the spin radius (R) is also constant. The objective of this analysis is to evaluate these five parameters given the aircraft physical characteristics. Although the emphasis in design is for spin avoidance and recovery it is important for aero engineers to understand the flight mechanics of aircraft spin.

Governing Equations

Spin is a stable flight condition in which all forces are in equilibrium. The aerodynamic moments are balanced by the gyroscopic moments similar to a spinning top. Any attempt to escape from this condition is countered by the dynamics to return to a stable equilibrium point. This situation makes it difficult to recover to a normal flight condition. In spin the power is off, the aircraft is stalled and the flight controls are ineffective. The governing equations shall be developed below.

It is convenient to use a cylindrical rotating coordinate system aligned with the earth. The cylinder has a radius of (R), an azimuthal angle (ψ) and a vertical axis emanating from the center of the earth. The unit vectors are $\hat{e}_1$, $\hat{e}_2$, and $\hat{e}_3$, respectively. The force equation for a rotating system is as follows:

$$T + W + A = m(\dot{V} + \omega \times V) \quad (1)$$

Where T = thrust = 0, W = weight = $W\hat{e}_3$, A = Aerodynamic Force = $L'\hat{e}_1 + Y'\hat{e}_2 - D'\hat{e}_3$, m = mass = W/g, and ω = angular spin velocity about the $\hat{e}_3$ axis. The prime values are used to designate the cylindrical system.

The velocity is composed of two components.

$$V = \dot{z}\hat{e}_3 + \omega \times R = \dot{z}\hat{e}_3 - \omega R \hat{e}_2 \quad (2)$$

Therefore

$$\omega \times V = \omega^2 R \hat{e}_1 \quad (3)$$

Since $\dot{V}=0$ for steady state, the governing equation for force contains three scalar equations:

$$L' = m\omega^2 R \qquad (4)$$

$$Y' = 0 \qquad (5)$$

$$D' = W. \qquad (6)$$

The centrifugal force is balanced by the inward radial lift force (L') and the weight is supported by the vertical upward aerodynamic drag force (D'). The tangential force (Y') is zero at equilibrium.

At stall, the aerodynamic force is nearly perpendicular to the wing surface and the axial force is negligible. Therefore,

$$L' = -N\sin\theta \text{ and } D' = N\cos\theta \qquad (7)$$

The aerodynamic coefficients are introduced next.

$$C_N = N/q^*S \qquad (8)$$

Where

$$q^* = 0.5\rho V^2 \qquad (9)$$

and S = planform area. Hence the three force equations become

$$\tan\theta = -\omega^2 R/g \qquad (10)$$

$$C_Y = 0 \qquad (11)$$

$$C_N\cos\theta = W/q^*S \qquad (12)$$

The moment equation for a rotating system is as follows.

$$M = \dot{H} + \omega \times H \qquad (13)$$

The only moment acting at the centroid is the aerodynamic term. The angular momentum is

$$H = I \bullet \omega \qquad (14)$$

where I is the moment of inertia tensor in 3 dimensions. It is best evaluated in the body axes system using principal axes which are known for any aircraft. Then since

$$\omega = p\hat{i} + q\hat{j} + r\hat{k} \quad (15)$$

$$H = pI_x\hat{i} + qI_y\hat{j} + rI_z\hat{k} \quad (16)$$

Hence

$$\omega \times H = \begin{vmatrix} \hat{i} & \hat{j} & \hat{k} \\ p & q & r \\ pIx & qIy & rIz \end{vmatrix}$$

$$= qr(I_z - I_y)\hat{i} + pr(I_x - I_z)\hat{j} + pq(I_y - I_x)\hat{k} \quad (17)$$

Also since $\dot{H} = 0$ and $M = L\hat{i} + M\hat{j} + N\hat{k}$. The moment equation in scalar form is the following:

$$\underline{L} = qr(I_z - I_y) \quad (18)$$

$$\underline{M} = pr(I_x - I_z) \quad (19)$$

$$\underline{N} = pq(I_y - I_x). \quad (20)$$

To convert from the body axes to the cylindrical axes a coordinate transformation is required. The Euler angles of pitch (θ) and roll(ϕ) are adopted.

Transformation Inverse   Inverse

$$\begin{vmatrix} \hat{e}1 \\ \hat{e}2 \\ \hat{e}3 \end{vmatrix} = \begin{vmatrix} \cos\theta & \sin\theta\sin\phi & \sin\theta\cos\phi \\ 0 & \cos\phi & -\sin\phi \\ -\sin\theta & \cos\theta\sin\phi & \cos\theta\cos\phi \end{vmatrix} \begin{vmatrix} \hat{i} \\ \hat{j} \\ \hat{k} \end{vmatrix} \qquad \begin{vmatrix} \hat{i} \\ \hat{j} \\ \hat{k} \end{vmatrix} = \begin{vmatrix} \cos\theta & 0 & -\sin\theta \\ \sin\theta\sin\phi & \cos\phi & \cos\theta\sin\phi \\ \sin\theta\cos\phi & -\sin\phi & \cos\phi \end{vmatrix} \begin{vmatrix} \hat{e}1 \\ \hat{e}2 \\ \hat{e}3 \end{vmatrix}$$

Since

$$\omega = \hat{e}_3 = p\hat{i} + q\hat{j} + r\hat{k} \quad (21)$$

Then

$$p = -\omega\sin\theta \quad (22)$$

$$q = \omega\cos\theta\sin\phi \quad (23)$$

and $r = \omega\cos\theta\cos\phi$. \quad (24)

Therefore

$$L = \omega^2 \sin\phi \cos\phi \cos^2\theta (I_z - I_y) = C_l q^* Sb \qquad (25)$$

$$M = \omega^2 \sin\theta \cos\theta \cos\phi (I_z - I_x) = C_m q^* Sc \qquad (26)$$

$$N = -\omega^2 \sin\theta \cos\theta \sin\phi (I_y - I_x) = C_n q^* Sb \qquad (27)$$

Next transform the moments from the body axes to the cylindrical system using the coordinate transformation.

$$M'_1 = M \cdot \hat{e}_1 = \omega^2 \cos\theta \sin\phi \cos\phi (I_z - I_y) = C'_l q^* Sb \qquad (28)$$

$$M'_2 = M \cdot \hat{e}_2 = \omega^2 \cos\theta \sin\theta (I_z \cos^2\phi + I_y \sin^2\phi - I_x) = C'_m q^* Sc \qquad (29)$$

$$M'_3 = M \cdot \hat{e}_3 = 0 = C'_n q^* Sb \qquad (30)$$

The moment coefficients were introduced above. The moment coefficients for the two axes systems are related as follows:

$$\cos\theta \, C'_l = C_l \qquad (31)$$

$$\cos\phi \, C'_m = C_m \qquad (32)$$

$$C'_n = -\sin\theta \, C_l + \cos\theta \sin\phi \, C_m c/b + \cos\theta \cos\phi \, C_n = 0 \qquad (33)$$

It is convenient to adopt the following definitions for the vehicle characteristics:

$$V_o^2 = 2W/(\rho S) \qquad (34)$$

$$\omega_o^2 = W c / (I_z \cos^2\phi + I_y \sin^2\phi - I_x) \qquad (35)$$

$$R_o = g/\omega_o^2 \qquad (36)$$

$$i_o = \omega_o^2 (I_z - I_y)/Wb \qquad (37)$$

$$\phi_o = \omega_o R_o / V_o \qquad (38)$$

These are the same similarity variables used to correlate model spin tunnel data with flight test. Using these definitions the governing equations become

$$-\tan\theta = \omega^2 R/g \tag{39}$$

$$\cos\theta C_N = V_o^2/V^2 \tag{40}$$

$$\phi_1 = C_l \tan\theta / C_m i_o \tag{41}$$

$$\text{or } \phi_2 = -C_n / C_m (c/b - i_o) \tag{42}$$

$$C_m / C_N = \omega^2 \cos^2\theta \sin\theta / \omega_o^2 \tag{43}$$

$$C_Y = 0 \tag{44}$$

$$C_n' = 0. \tag{45}$$

Since the aerodynamic coefficients depend upon the angle of attack (α) and the sideslip angle (β), it is necessary also to express them in cylindrical axes.

Since

$$V = \dot{z}\hat{e}_3 - \omega R \hat{e}_2 = u\hat{i} + v\hat{j} + w\hat{k} \tag{46}$$

Hence

$$u = V \cdot \hat{i} = -\dot{z}\sin\theta \tag{47}$$

$$v = V \cdot \hat{j} = \dot{z}\cos\theta\sin\phi - \omega R\cos\phi \tag{48}$$

$$w = V \cdot \hat{k} = \dot{z}\cos\theta\cos\phi + \omega R\sin\phi \tag{49}$$

Definitions

$$\tan\alpha = w/u = -\cos\phi\cot\theta - \omega R\sin\phi/\dot{z}\sin\theta \tag{50}$$

$$\sin\beta = v/V = \dot{z}\cos\theta\sin\phi/V - \omega R\cos\phi/V \tag{51}$$

Approximate Solution

To evaluate the five spin parameters it is necessary to know three vehicle dynamic quantities which are V_o, ω_o and i_o. In addition four aerodynamic coefficients C_N, C_l, C_m, and C_n must be related to the flight conditions. To obtain an understanding of spin the following approximations simplify the governing equations. The terms α and θ are large while β, ϕ, and $\frac{\omega R}{V}$ are small. The aerodynamic force during stall is nearly perpendicular to the wing surface and the center of pressure moves from about the quarter chord to the midchord position to produce a negative pitch down moment coefficient. Wind up into spin proceeds until the side slip angle nears zero. The above simplified equations generate five universal relationships for aircraft spin as a function of pitch angle.

$$V^2/V_o^2 = 1/C_N \cos\theta \qquad (52)$$

$$\omega^2/\omega_o^2 = C_m/C_N \sin\theta \cos^2\theta \qquad (53)$$

$$\phi_1 = C_l \tan\theta/C_m \, i_o \qquad (54)$$

$$\text{or } \phi_2 = -C_n / C_m (c/b - i_o) \qquad (55)$$

$$R/R_o = -C_N \sin^2\theta \cos\theta/C_m \qquad (56)$$

$$C_n' = -C_l \sin\theta + \phi \, C_m \, c/b \, \cos\theta + C_n \cos\theta = 0 \qquad (57)$$

By combining the last three equations a simple expression for the spinning moment may be derived.

$$C_n' = (\phi_1 - \phi_2) \, C_m(c/2b - i_o) \cos\theta = 0. \qquad (58)$$

It is obvious that equilibrium occurs when $\phi_1 = \phi_2$. In addition, simple relations may be deduced for the orientation of the wind axes.

$$\tan\alpha = -\cot\theta \text{ or } \theta = \alpha - 90 \text{ and } \beta = \phi\cos\theta - \omega R/V. \qquad (59)$$

Given the four aerodynamic coefficients and three vehicle characteristics, the five spin variables may be expressed as a function of only pitch angle. The roots of C_n' equal zero. This equation will determine the appropriate spin pitch angles.

Solving Procedure

To demonstrate the procedure, the spin conditions were predicted for three fighter aircraft described in a NASA report (ref 4). In this report the dynamic similarity conditions (V_o, ω_o and i_o) were given for each aircraft and the corresponding aerodynamic coefficients were tabulated for the entire angle of attack range (θ). The three fighters described in the report are as follows:

Configuration (A) variable swept wing, F-111

Configuration (B) delta wing, F-106

Configuration (C) swept wing, F-4

Table 1

	V_o (Feet/sec)	R_o (Feet)	i_o	ω_o (Rad/sec)	ϕ_o
Config. A	463	20.280	0.020	1.260	0.055
Config. B	283	6.776	0.050	2.180	0.052
Config. C	381	8.468	0.028	1.950	0.043

Calculations of the five spin variables (V, ω, R, ϕ and C_n') were accomplished as a function of pitch angle. All calculations were performed for a spin altitude of 30,000ft. The proper pitch angle was determined for the roots of the spinning moment equation (C_n'). Multiple roots were found for all aircraft. A stability analysis shows that $dC_n'/d\omega$ must be negative for the stable spin. The extraneous roots were eliminated to only the stable ones. In all cases two stable roots were obtained which are flat and steep spin. Plots of the spinning rate versus pitch angle and spinning moment versus pitch angle are presented in Figures 1-6 for the three aircraft. A tabulation of the computer results is shown in Table 2 for the three aircraft. A summary of the stable equilibrium points for the aircraft is shown below.

Table 2

	Configuration	θ	V (Feet/sec)	ω (Rad/sec)	R (Feet)	φ	β
A	Flat	22	298	1.48	6.2	1.1	-1.6
A	Steep	68	578	1.57	34.1	-3.9	-6
B	Flat	23	257	1.23	9.3	0.4	-1.9
B	Steep	57	337	1.13	38.9	-4	-9.6
C	Flat	29	309	1.89	5.1	-2.2	-3.6
C	Steep	69	621	2.1	19.1	-8.4	-4.4

All angles are in degrees.

An involved numerical procedure was used in Ref. 5 to determine the spin characteristics for configuration B. The results were as follows in Table 3:

Table 3

	Configuration	θ	V (Feet/sec)	ω (Rad/sec)	φ	β
B	Flat	24	261	1.17	0.9	-1.9
B	Steep	55	326	1.1	-4	-9.6

Acceptable agreement with our results is observed for both cases demonstrating the success of the simple procedure.

Conclusion

In this paper a tutorial is presented to provide a simple analysis for the understanding of aircraft spin, i.e. a stable equilibrium condition in which the centrifugal force is balanced by aerodynamic lift and the gyroscopic moment is balanced by the pitch down moment. Two stable equilibruim roots were found for three aircraft using this simple procedure.

After determining the characteristics of stable equilibruim spin one is tempted to explore recovery techniques. To extricate from spin it is difficult to alter the piching moment due to the ineffectiveness of the stalled elevators and due to the huge moment caused by the large shift of the normal force. However the small lateral moments may be changed easily to interrupt the stable spin state. Some nose device could be used to disturb the nose tip vortices and produce recovery.

References

1. A. Glauert "Investigation of the Spin of an Aeroplane", ARC R & M No. 618 , June 1919

2. R. Woodcock, T .Cord "Stall/ Spin : Seventy Years Later" Air University Review May 1974

3. K. Wurtzler, "F-15 Spin Analysis" Gridgen Cobalt Solutions 2010

4. W.Gilbert, C. Libbey, "Spin-Prevention System for Fighter Airplanes" NASA TND-6670 , 1972

5. D. Westgate "Analytical Determination of Aircraft Spin Modes" AFIT Thesis , 1974

Appendix

The arrows in Figures 2, 4, and 6 point to the stable equilibrium spin conditions.

Figure 1

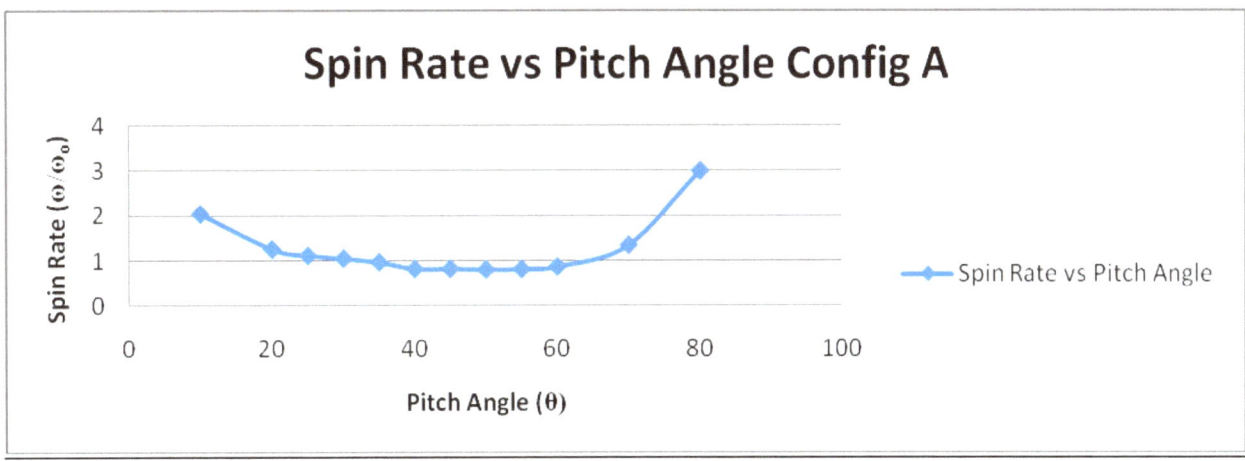

Figure 2

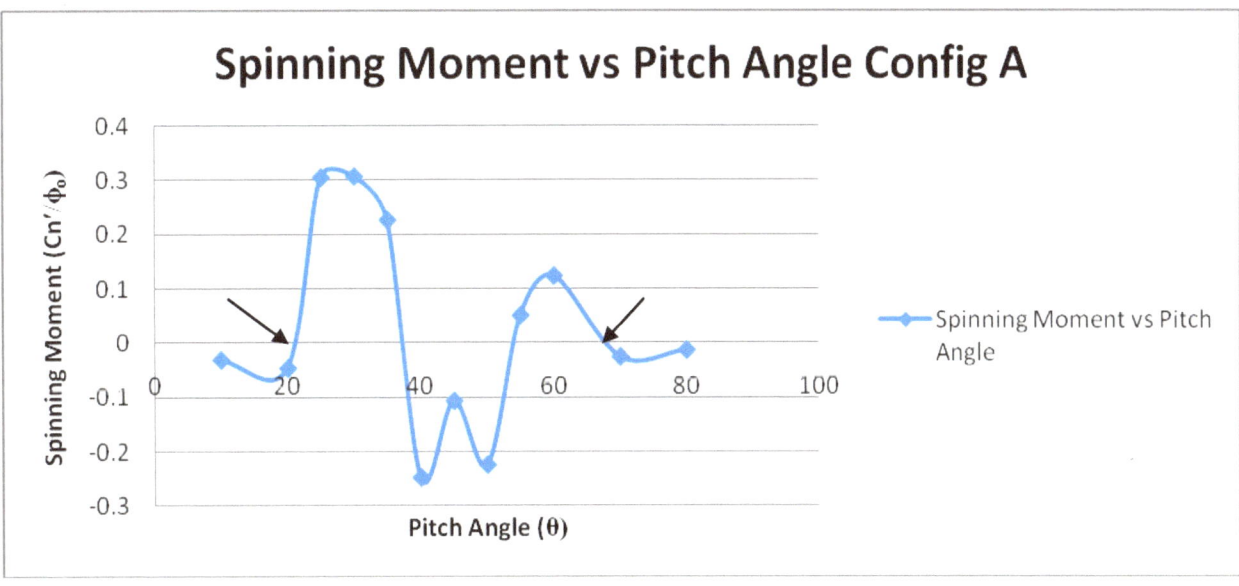

Figure 3

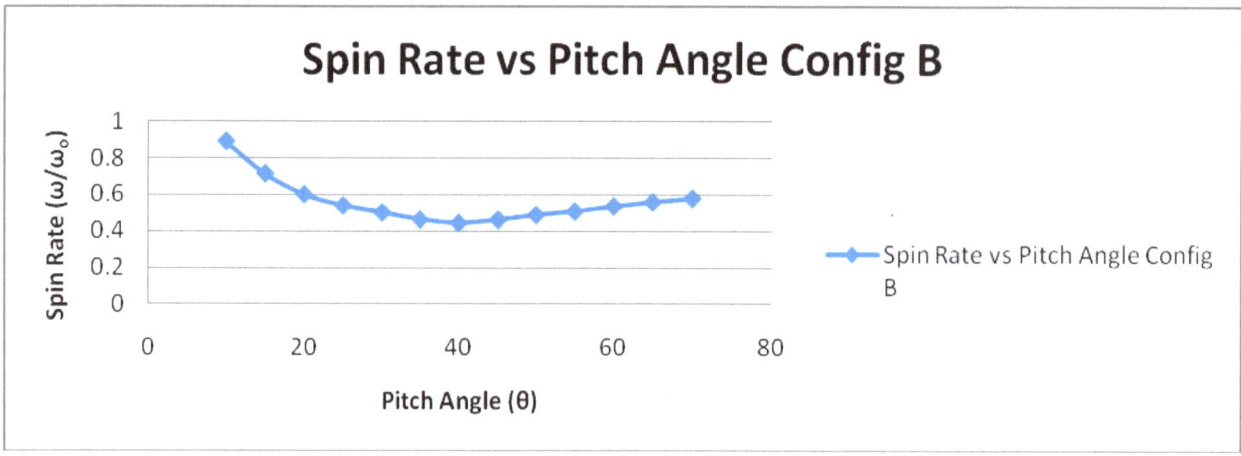

Figure 4

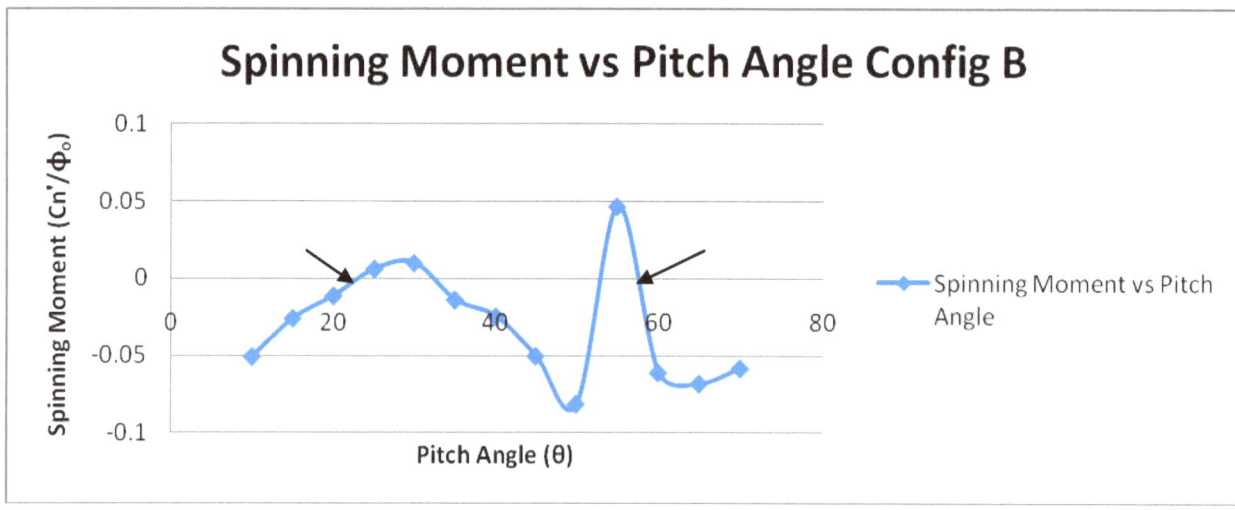

Figure 5

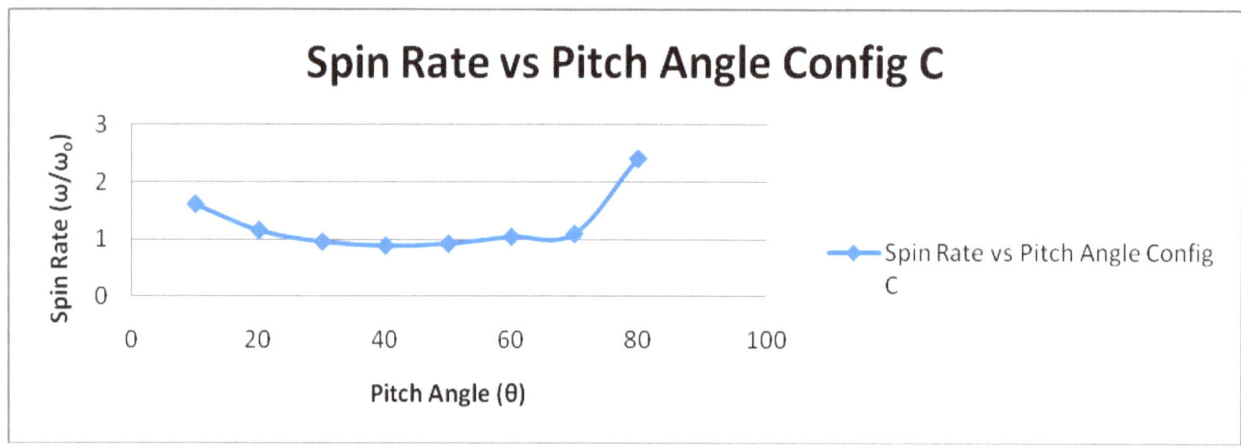

Figure 6

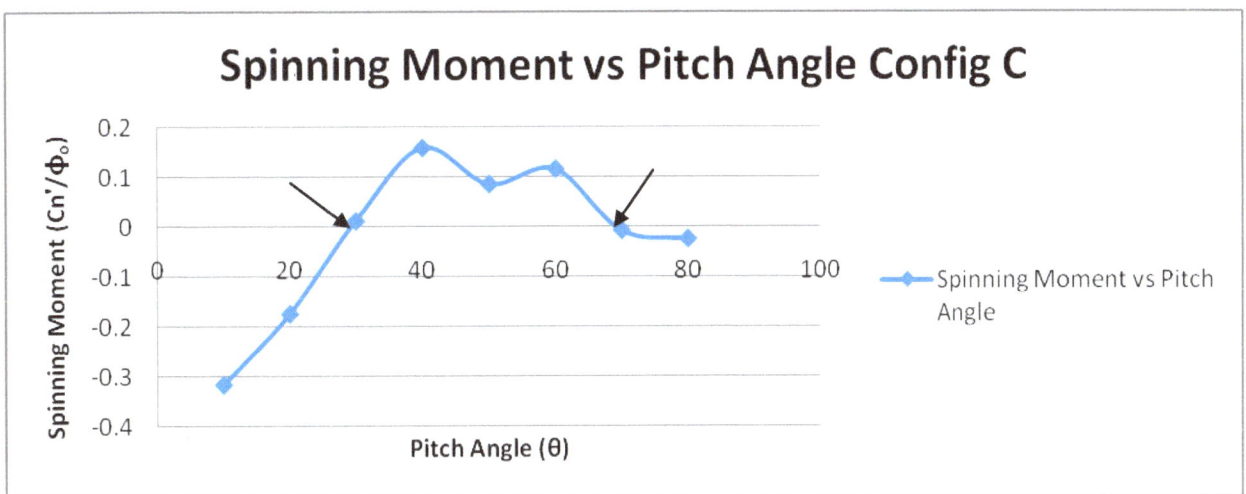

www.ingramcontent.com/pod-product-compliance
Lightning Source LLC
Chambersburg PA
CBHW050438180526
45159CB00006B/2581